QUANTUM PHYSICS

for Beginners

The Layman's Guide to Understand How Everything Works.
Look Into the **Mind-Blowing Secrets** of Science
in a Comprehensible Way, From **String Theory** to **Quantum Computing**

Gage Dotson

TABLE OF CONTENTS

INTRODUCTION

The Other Side of Reality —
Welcome to the World of Quantum Physics

There is this belief that quantum physics isn't for everyone. While that might be true, it isn't totally right. What I mean is that anyone can understand quantum physics with enough time, dedication, and the right material. Speaking of the right material, I assure you that you are truly in the right place if you are reading this book.

You don't have to worry about scary formulae and technical jargon that always leaves you shaking your head and just giving up—this book contains little of it. Instead, it focuses on explaining quantum physics' basic concepts; not only in words you can understand but in ways you can follow.

Our goal is to help you understand quantum physics in the easiest way possible. We will look at how the smallest pieces of our environment and the universe affect our lives and the scientists that came up with the science behind quantum physics. And we will do all these without giving you those headaches associated with understanding the concepts around this branch of knowledge.

We are not saying there won't be any math, but we can say it will be painless. So just enjoy the ride as we take you through the science behind atoms and the universe as a whole.

One of our primary focuses will be on some of the great minds who sacrificed their time in the laboratory carrying out experiments that contributed to our modern-day understanding of quantum physics. We will also look at the flaws in some of the theories founded by these minds and how they contributed to other, better experiments.

The study of quantum physics aims to explain how the tiniest pieces of the universe contribute to the big picture we now have as our world. When humans finally gain the ability to travel to other worlds and live there, quantum physics will be one of the major contributors to that happening. The same could be said when we can finally travel through time. If those things are not your goal, but you want humans to gain the ability to teleport from one place to another and finish errands quicker, then it all comes down to quantum physics.

These things may seem like something out of a sci-fi movie. However, many of them are already happening at a practical level. You can keep your fingers crossed that you will see something big happen one day. What we are trying to say is that there is really nothing to be afraid of when it comes to quantum physics.

As you read through everything in this book, you will encounter several scientists that contributed to today's quantum physics. You will realize that even though they are not completely normal (Schrödinger is definitely at the top of the list of weird ones), they are like you in many ways. They are humans like you who want to understand their environment and why the universe is the way it is. To do this, they had to look at the smallest unit of our existence that makes up everything. And that is why we are going to start with the beginning: the atomic theories.

What Is Quantum Physics?

To truly understand quantum physics, you have to go back to where it all began. In other words, physics. Yes, I know; physics is not the most interesting subject for many people. To like physics in school means you need to be a nerd, and many of us don't want that. Surprisingly, many students even find mathematics and chemistry easier to understand than physics.

Lots of people find physics boring because it involves mechanics; in other words, "the principles governing how things work." Although this is true, physics goes beyond just how things work.

The origin of the word physics is interesting indeed. It comes from *physique*, a Greek word meaning "knowledge of nature." This means that physics helps us to understand the science and principles governing nature. Another definition of physics is the study of matter, how it responds in space and time, and its connection to forces and energies.

Each aspect of these concentrates on the knowledge of matter and quantities. An example is seen in nuclear physics which deals with the study of atomic weight and how it responds in different environments.

Another way to consider physics is in terms of classical and modern physics. Although quantum physics is unique, it is sometimes used synonymously with modern physics. So, while it is a continuation of classical physics, it is also its opposite too. While most of what happens in contemporary physics is based on quantum theory, quantum physics is still a branch of modern physics.

Quantum physics lies at the core of everything in existence, and by reading this book, you can join the billions of people who are on a journey to understand the nature of existence at the smallest level.

CHAPTER 1: WHAT'S THE MATTER?

A Short Deep Dive Into What Makes Matter a 'Thing'

Lab equipment like mass spectrometers and electron microscopes weren't just discovered in one day. Our understanding of quantum physics today is the result of numerous experiments and discoveries by many minds. The story of the origin of quantum physics has its basis in classical physics. Therefore, we are going to begin explaining quantum physics from classical physics.

Classical physics is a branch of science that tries to understand how mechanics and mathematics work. There is no clear line or date that demarcates the end of classical physics and the beginning of quantum physics. The concepts and theories that evolved into quantum experiments began slowly and gained huge momentum during the 1900s into what it is today. The first human experiment in quantum physics began as a result of human curiosity, which evolved into another branch of physics. It began because scientists couldn't understand many issues with the laws of classical physics; so, they needed answers. This led to the birth of quantum mechanics and other branches of physics that deal with the quantum world.

Quantum physics was already in existence even before the apple drop that led to the origin of classical physics. The Greek philosopher Democritus was the first person to introduce the earliest recorded concept of physics in 400 BCE. He came up with what is known as the theory of the universe, which can be summarized in the following principles:

- All matter is made of atoms that cannot be seen with the naked eye.

- Empty space exists between atoms.

- Atoms are solid mass, but they do not possess a set internal structure.

- Substances are made of different atoms, each with its own weight, size, and shape.

Although Democritus' theory is more or less generalized, it is in line with our understanding of matter and atoms to some extent. The most surprising point is that he arrived at his theory with only philosophical thought experiments rather than scientific experiments. According to Democritus, every substance can be continuously divided until it is no longer divisible, and the smallest quantity of indivisible matter is the atom. His theory played a huge role in our early understanding of matter and remained relevant until the 19th century.

The Atom and the Quantum Physics Model

Early Atomic Models

The turn of the 1800s saw John Dalton become the first modern scientist to carry out research based on the atomic theory proposed by Democritus. Dalton's atomic theory resulted from his effort to understand what Democritus really meant. His theory included many ideas of Democritus, but it also incorporated something new: Dalton introduced the world to the concept of atomic weights, which can differentiate between substances.

Dalton proposed the following statements in his atomic theory:

- All matter is made up of tiny particles known as atoms.

- Atoms can't be destroyed or further divided.

- Atoms of the same elements have identical weights.

- Atoms of different elements combine to form compounds. In a compound, the atoms are bonded in constant proportions (if the proportion of atoms in a compound is fixed, so is the proportion of their masses).

Although Dalton's atomic model was more precise than that of Democritus, it received some backlash, mostly due to accusations by the Irish chemist William Higgins. According to Higgins, Dalton's atomic model was stolen from his work.

But it happened that Higgins was wrong. Funnily enough, Dalton's idea for his atomic model was taken from a chemist named Bryan Higgins, who turned out to be William Higgins's uncle. The situation must have been hilarious because Dalton had to defend himself by saying that his idea was not stolen from Higgins but from his uncle.

Irrespective of the controversy, Dalton's atomic theory was really his own, even if he based his idea on Democritus's theory and findings from Higgins's uncle. Notebooks belonging to Dalton showed that his experiments into matter and atoms were unique and that some notions predicted the relative weights of certain elements. Even though many of the concepts were not his, he came up with new ways to study these topics. And for this, he is credited with being one of the founders of the atomic model. His papers were kept in the city of Manchester, England. However, many were lost or destroyed in 1940.

British physicist J.J. Thomson picked up from where Dalton left off and proposed his "Plum Pudding Model" of the atom. His major contribution to the modern-day atomic model is his discovery of the electron with the aid of cathode rays. After finding the electron, he released a revised version of his atomic model. Then, the next model of the atom he proposed was even further away from the truth. He only discovered that electrons existed; however, he failed to predict how they worked.

Thomson's model didn't have much impact on the atomic theory that already existed. His concept was a little crooked since it showed the presence of negatively charged electrons dominating the atom. This was what gave rise to his notion being called the plum model. Although his failed atomic model dominates Thomson's work, his discovery of the negatively charged electrons was significant in getting quantum physics to where it is today.

Several scientists were not satisfied with what Thomson put forward, and one of those people was Ernest Rutherford, popularly credited with being the grandfather of nuclear physics. He was highly respected among his peers and was friends with many brilliant minds, including Thomson, during his time. He was troubled by Thomson's atom model, so he decided to work on discovering what was wrong with it.

His answer to what is wrong with Thomson's model arrived while he was at the University of Manchester, England, coordinating the Geiger-Marsden gold foil experiments. The famous experiments were carried out by using a stream of alpha particles to bombard a thin piece of gold sheet. The experiment's goal was to observe any reaction or changes in the atoms present in the stream of alpha particles. To the surprise of the people conducting the experiment, they noticed that not all the alpha particles were going through the gold foil. Rather, many were being deflected at various angles, some at an angle of even 180 degrees. You may be wondering how this was important and how it contributed to Rutherford's atomic model.

From his experiment, Rutherford discovered that the atom isn't just a gelatinous blob as proposed by Thomson's model. He deduced that helium atoms must possess a hard center or something similar that prevented them from completely going through the gold. And his discovery was right! What he found was that all atoms possess a nucleus. With this, the scientific world said goodbye to Thompson's plum pudding atomic model and welcomed Rutherford's atomic model. The model proposed by Rutherford is an atom with a central nucleus containing positively charged particles called protons that are held so close that they can't go through another element and negative-charged electrons surrounding the nucleus at orbits. This model persisted for only two years before the arrival of the Rutherford-Bohr atomic model or just Bohr's atomic model. Despite this, Rutherford was still given the Nobel Prize, was made a knight, and received a barony in recognition of his outstanding discoveries.

The atomic model proposed by Bohr was the first model showing electrons surrounding the nucleus in orbits, which brought scientists closer to what the atom really contains. You might have learned about this model in school at some point, although it isn't completely accurate.

The most accepted model in modern-day physics was suggested in 1926 by Erwin Schrödinger (we are going to see more about this phenomenal mind later). He proposed a model that described the atom as having an electron cloud where electrons fluctuate in waves towards and away from the nucleus rather than in set orbits suggested by Bohr. Schrödinger got the idea for his model from the wave-particle duality theory, which we will look at later.

For now, we will focus on the atomic models that began with Democritus, Dalton, J.J. Thomson, Rutherford, and Bohr.

The Double-Slit Experiment

The double-slit experiment, realized in 1801 by Thomas Young, also recognized as the origin of the wave-particle duality, was a major contributor to the move from classical physics to quantum physics. This experiment was performed by sending an energy-concentrated beam of light, like a laser, to a flat solid surface with two narrow slits made on it. The energy is projected through both slits onto a flat surface. You may ask what the point of this experiment is. This experiment shows that light acts as both a particle and a wave. This is the most basic principle of wave-particle duality. The discovery of this phenomenon is an important foundation in quantum theory. When light passes through just a slit, it appears behind as a single bar of light or illumination. However, when it is made to pass through double slits, the light breaks up and appears behind in a striated pattern. Light behaves this way because of the wave-particle duality.

Before this experiment was carried out in 1801, scientists believed that light behaved either as a particle or a wave. After it was proven that light could behave as both, a lot of work was put into researching the dual-wave theory.

Scientists also believed that the only thing in nature that could behave in this way was light, but they discovered in 1927 that electrons also behave according to wave duality. After that event, scientists also found out that all atomic particles could behave the same way.

You may ask, what is wave-particle duality and what is its significance to quantum theory? First, we are going to define waves according to science. A wave refers to a disturbance in a medium that causes energy movement. Waves include sound waves, which cause disturbances in the air and water, and radio waves which move through the atmosphere.

The properties that determine the type of energy a wave is transmitting include amplitude, the wave's height, and frequency, which is the speed the waves take to go through a whole wavelength. Both classical and quantum physics concentrate on waves in the electromagnetic spectrum. Different scientists made new discoveries for over 100 years that contributed to our understanding of what the spectrum is, and most of these discoveries are still relevant today. The electromagnetic scale is arranged starting with the wave with the highest frequency and continues until it gets to the wave with the lowest frequency. We have:

- Gamma rays

- X-rays

- Ultraviolet light

- Visible light (violet, indigo, blue, green, yellow, orange, red)

- Infrared light

- Microwaves

- Radio waves

After the spectrum was completed, scientists were still left with several questions. Why can we see light, which travels through a vacuum, if waves can only travel in a medium? As it turns out, electromagnetic waves are special in that they can travel even in the absence of a medium. This revelation was one of the points that began to separate quantum physics from classical physics. Many scientists were left with many questions, such as "Can we classify electromagnetic waves such as light as waves since they can travel in vacuum or particles?" The answer they came up with was that light is both a wave and a particle.

If you are having a hard time understanding this phenomenon, don't worry. As you read on, it will begin to sink in. After all, people receive salaries to think about it, and they can't completely understand it. You need to know that the smallest units of the universe behave both as waves and particles. Isn't that cool?

Cathode Rays and Black-Body Radiation

We will go into phenomena that were once only possible in sci-fi stories and our imagination but are now being seen in laboratories. J.J. Thomson didn't just wake up one morning to find electrons scattered all over his room; he had to dedicate time working on a way he could prove their existence with cathode rays because, in the world of science, the proof is all that matters.

A cathode-ray tube is a piece of equipment made of a glass tube, with its inner gas removed and sealed. This creates a vacuum (or an almost vacuum) in the tube. One end of the glass contains a metal cathode that is electrified with a high charge. This charge is directed through an anode into a cathode ray. A positively charged metal plate with a magnet is placed on one side of the tube. The same is done for the opposite side but with a negative charge. During the experiment, scientists could determine the charge and movement of the atoms present in the tube as the cathode ray moved in the tube. J.J. Thomson discovered the existence of electrons during one such cathode ray experiment. He found out that the atoms in the tube behaved in an odd way that was not expected. He performed further experiments and concluded that something must be responsible for the fluctuations he noticed in the weight and behavior of the atoms he was measuring.

Black-body radiation is another phenomenon that dominated the 20th century. Although the name might make it sound spooky, you will understand those parts of our universe that we can't see once you read the explanation below. Another name for black-body radiation is thermal radiation. The theory of black-body radiation helps us to understand how energy equilibrium is preserved throughout all matter in the universe. A black hole is a classic example of a measure of black-body radiation that is unbalanced. A black hole absorbs all energy near it, irrespective of its wavelength. With the aid of the black body, radiation scientists can detect the presence of matter which they can't identify just by measuring the thermal energy of waves in the surrounding. Science defines a black body as matter that absorbs all electromagnetic energy it comes into contact with and radiates it by following the laws of thermodynamics.

To understand the meaning of a black body, think about a parking lot that receives lots of heat from the summer sun during the day. As the sun shines during the day, the parking lot absorbs energy in the form of heat and light. When it's night, the parking lot then releases this energy back into the atmosphere to reach equilibrium. A black body behaves in this way. Ignoring the size and appetites of black holes, scientists can predict the size and magnitude of the black bodies in their labs and space by simply tracking the motion of electromagnetic waves and looking out for where they disappear and reappear.

Radioactivity

Radioactive waves like gamma rays have several benefits ranging from a useful energy source to health benefits. They can also lead to tissue destruction because of their ability to destroy human life. Lead shields are the only effective means of protection in X-ray procedures. For this reason, Chornobyl, the site of a famous nuclear accident in 1986, will remain uninhabited for a long time. The disaster occurred on April 26, 1986, at its nuclear power plant, located in present-day Ukraine. During a safety test, an explosion occurred in reactor number 4, releasing a large amount of radioactive material into the air and contaminating much of Europe. The catastrophe caused the immediate death of two plant workers and a firefighter, and it is estimated that, in the long term, radiation caused the deaths of thousands of people. The calamity also had significant environmental and economic consequences and led to an increased focus on nuclear safety worldwide.

William Roentgen discovered that you could see into the human body with the aid of X-rays in the 1800s. Someone who was fascinated by his discovery was one of his colleagues back then known as Henri Becquerel, a third-generation French scientist. Becquerel began to think that maybe X-rays could be used to explain what he observed in his experiments with phosphorescent minerals and elements. He wasn't just known for being a scientist. He was also good at taking pictures because of his photography skills. Becquerel came up with a method to determine if the phosphorescent minerals he was working with were absorbing light and then radiating it back as X-rays. The experiment was pretty simple. All he had to do was to expose the minerals to light. After that, the minerals are then between two photographic plates covered with black paper. If the minerals were radiating X-rays, then they should show on the plates just like photographs. The experiment worked at first, and to be sure, Becquerel did the same experiment exposing the minerals many times on sunny days, and he was able to confirm the appearance of uranium salt on the photo plates. He also went ahead to add a few solid objects such as coins for control, aiming to ensure that what he was seeing weren't just dust particles.

Becquerel was sure that his phosphorescent mineral was indeed radiating X-rays. He then published academic papers about his findings and submitted them to some of his colleagues. However, he continued playing with his experiment to capture pictures of the X-rays themselves. He continued this process until one day it stopped working, and he wasn't able to get the lighting effect he desired. Like any good scientist, he threw his research away and continued with his usual life. After a few days, probably out of scientific curiosity, he took the photo plates for development even though they hadn't been exposed to sunlight. The result was that he obtained the clearest images he had captured since the experiment. How was this possible when he didn't expose the uranium salts to the sun before placing them between the photo plates?

The only reasonable explanation is that the uranium salt was emitting its own radiation, a type that doesn't depend on light. This came as a huge surprise to Becquerel. He carried out the uncalculated experiment several times, each time with different uranium compounds, to find out if the images he got were a result of leftover phosphorescence or if they were from X-rays. The results he got were all the same. He had unintentionally discovered radioactivity.

Pierre Curie and his wife Marie became fascinated by Becquerel's experiments while doing their own experiments with uranium. Marie was the first to come up with the term "radioactivity." She was also collateral for the couple's work with uranium and other radioactive elements. Polonium and radium were both Marie's discoveries. Thanks to radioactivity, humans have been able to achieve great things, such as nuclear power and the capacity to date old objects. But it has also caused numerous tragedies. This is because the atoms of radioactive elements can emit their own radiation while breaking down. This feature is based on the atom's unstable nature.

Becquerel and the Curies received a Nobel prize for their pioneering experiments on radioactive elements. Marie also became the first woman to receive two Nobel Prizes, but her achievements came at the cost of her life. Although Pierre Curie died in 1906 due to a road accident, Marie continued their work together, creating mobile X-ray units to be used in ambulances in the First World War and accurate theories on several radioactive elements. She also did all these while raising two daughters. Her price for excessive exposure to radioactivity was blood cancer, but she continued her work, pushing the limits of scientific advancement, until the day she died in 1934.

Her casket was lined with lead to keep in the radiation that would continue to emit from her remains for the next 1200 to 1500 years. The death of Becquerel was also due to complications from radioactive poisoning. The work of both Curies and Becquerel with radioactivity tells us all we need to know about the nature of their experiments. The price they paid for their discoveries does take some shine from their achievements, but if Becquerel didn't accidentally discover that his uranium salts were emitting radiation on their own, the world would have taken some time to pinpoint radioactivity. This shows the danger that the early quantum physics scientist faced while diving into unknown territory. Two of the three pioneering scientists of radioactivity died due to their work with what they discovered. Maybe even Pierre Curie would have gone on to do great things in quantum physics.

The Franck-Hertz Experiment

The experiment carried out by Gustav Hertz, and James Franck in 1914 helped confirm several energy states of atoms. During the experiment, the two scientists facilitated the movement of electrons in an electron tube containing gas. The electrons reached their specific binding energy as their energy increased. The electrons then stopped moving abruptly after their constant journey in the tube. When electrons achieve their specific binding energy, they expend it when they collide with nuclear electrons and move to a higher power level.

Lights, Camera, Inter-Action!

What Light Really Is and How We See Things

We interact with things when light bounces off objects and then hits our perceptual organs and brain. But light cannot "touch" particles. You wouldn't see anything if the light didn't exist, the same way you can't see in a dark room. The complicated interaction of light with our eyes and brain enables us to see. Light moves through space to enter our eyes, which causes a signal to be sent to the brain. The brain then interprets the information, which helps you to determine the location, appearance, and movement of what you are seeing. Light is the most important factor for this process to happen.

Luminous Versus Illuminated Objects

The things we can see can be grouped into two classes: luminous and illuminated objects. Luminous objects can produce their own light, while illuminated objects can't produce light; they only reflect it. The sun is an example of a luminous object, while the Moon and other objects around you are illuminated.

The light from luminous objects like the sun enables us to see our surroundings: the blue sky, the green grass, cars, and so on. We can see these objects because they reflect the rays of light into our eyes which our brains process as images. When the sun goes down in the evening, objects and our surroundings appear dark because of the absence of light.

One experiment that is carried out to determine the importance of light to sight is making a laser beam travel across a room. The lights in the room are turned off before turning on and directing the beams to the plane mirror. The beam travels to the mirror before reflecting the opposite wall at an angle. The light beam can't be seen when traveling to the mirror and after reflecting on the mirror.

The only places where you can detect the light are at the point where it strikes the mirror and where it strikes the opposite wall. You can only detect the beam at these points because the mirror and wall reflect it into your eyes. For you to detect an object, light from the object must be reflected into your eyes, and there are just two places where the light reflects. The beam in between the source and the mirror can't be detected, and the same can be said about the light reflected from the mirror to the wall.

During this experiment, small water droplets are seen if water is sprayed in the region where the light beam moves to the mirror. This is because these droplets reflect some portion of the light toward your eyes. The presence of water droplets helps us see the path of the light beam as it travels. The beam's light strikes the water droplets and is reflected toward your eyes. This just solidifies the fact that we can't see without light.

Like other illuminated objects, humans cannot produce light in the visible part of the electromagnetic spectrum. We are not brilliant like the Sun that shines so bright (no offense!). We are much more like the moon that reflects light. Other people can only see you because you reflect light entering their eyes. We can only see most objects in our environment because they reflect light, not because they produce light.

CHAPTER 2:
THE ORIGIN OF QUANTUM PHYSICS

To understand how quantum physics describes matter, we will look at the beginning of how this branch of physics came about. Every object in the universe is constantly exchanging energy. This energy is called electromagnetic radiation. As the temperature increases, the spectrum of colors we can see gradually changes to white. As we gradually move away from the visible light spectrum, we get into the spectrum of ultraviolet waves. As we move higher in the electromagnetic spectrum, the wavelength becomes shorter and shorter. In summary, as the temperature of an object increases, the range of radiation gradually moves up from the infrared range to the ultraviolet range.

The questions you may ask are: What is causing this gradual change? Why is it that these objects don't just emit their entire radiation equally? We are going to go off track a bit into the divergence between reality and theory, which is referred to as the ultraviolet catastrophe. Going with this theory, objects are supposed to emit all the energy they have at once because they don't have unlimited energy. In reality, this will mean that the world should be covered in radiation, specifically gamma radiation and very cold objects, which should take longer to heat.

It gets even more complex if you attempt to accurately measure only the radiation emitted, differentiating it from the radiation being reflected. But the problem is that no one has found a perfectly black body. However, scientists have invented bodies that reflect very little radiation, which allowed them to carry out experiments with more accurate results. This has enabled scientists to determine the precise radiation emitted by objects depending on their temperature.

Another issue that scientists were trying to solve was to explain why the radiation emitted by objects is limited to a certain range and quantity. This mystery was a huge problem for scientists, especially since it deviated from already-established laws that proposed that it is a must for objects to emit all their energy in the form of radiation with unlimited intensity. These laws have already explained several important issues, so it baffled scientists that they couldn't come up with an explanation. When the revolutionary solution to the issue finally presented itself, it was shunned by the majority of the scientific community and was rejected. What was even worse was the fact that the person who came up with the solution began to doubt himself. The radiation emitted by objects is limited to a certain range and quantity due to energy quantization. Objects emit energy in the form of photons, which are elementary particles of light, and the amount of energy contained in each photon is directly proportional to its frequency. This means that the radiation emitted by an object depends on the amount of energy available in the object and the frequency of the radiation.

The solution to the mystery of quantization of the energy and radiation emitted by objects was discovered by Max Planck at the end of the 19th century. Planck had attempted to explain the emission of radiation by hot bodies and discovered that the energy was not emitted continuously, as might be expected, but in discrete packets or quanta of energy. This discovery led to the birth of quantum physics and marked a break with classical physics, which did not contemplate the quantification of energy.

The general rule known to everyone, including you, is that fractions are used to express any quantity in science, ranging from weight to length. If a whole number like 5 is not accurate enough to describe a given quantity, then a fraction can be used to increase the precision of the quantity. For instance, we can have 5.5, 5.5423, 5.4234221113, and more. When it comes to describing a value, a minimum number doesn't exist since the value can be fractional and made smaller and smaller. This means that 0.1 is greater than 0.01, which is then greater than 0.001, and so on. This idea governed everything to do with energy in physics until it fell short in the 20th century. In other words, to fully understand the mathematical theory of radiation proposed by Max Planck (Planck's law), the idea had to be modified.

According to Planck, the goal of his law isn't to provide a complete solution to the fact that heated objects emitted only a limited amount of radiation. His law was to provide a suitable mathematical scenario that could explain the radiation emitted by objects to a reasonable extent. His goal was to create realistic calculations to explain emitted radiation of objects at high temperatures. Planck's law is summed up in the following logic below:

The value for the lowest possible amount of radiation emitted is one quantum (this value depends on the wavelength). This means that a value of an emitted radiation cannot be a fraction or ½ quantum. As a consequence, the number of radiation-emitted values can only be recorded in whole numbers of quanta and not fractions. If the temperature of a group of particles is low, then it simply means that it has a low amount of energy. When an object has low energy, the atoms present in the object will produce a smaller number of quanta or radiation with a longer wavelength than articles with higher energy levels. If the atoms in the object don't possess enough energy to make one quantum of ultraviolet X-rays or gamma-rays, it won't emit any radiation. So, your everyday objects with ordinary temperature can't produce any quanta of light. Instead, they emit lower-energy quanta with low frequency, i.e., long wavelengths such as microwave, radio wave, and infrared.

The most unique characteristic of Planck's model is that, for unknown reasons, energy can only be emitted and absorbed in certain quantities. The dose of emitted or absorbed radiation is measured within a certain value of these radiation units. In the same way, any given amount of silver will have a certain number of atoms; radiation is made of quanta (energy particles) with a size that is specific only to that given wavelength or frequency.

In summary, according to Planck, shorter waves with higher frequencies have larger quantum, which reduces the radiation from these shorter waves. He also suggested that the amount of energy for each quantum of radiation is always directly proportional to the frequency of the particular radiation. For example, suppose you want to calculate the power in a given radiation quantum of a particular rate. In that case, all you have to do is multiply the frequency of the wave by the Planck constant, also known as the standard coefficient.

With Planck's formulas, scientists could predict and describe the precise radiation of hot or heated objects instead of an approximate value. Planck's method of dividing the energy quantities into quanta also provided solutions for several other issues in physics. This splitting of the energy quantities is called quantization. Because of quantization, quantities can only gain certain specific total values such as 1 or 109 quanta of energy.

This simple concept of quantization proposed by Planck revolutionized the world of science. Quantum physics applying Planck's idea is the most successful and effective theory made by a human. Almost everything in modern physics today has benefited from the concept. Forms of modern, sophisticated technology apply Planck's concept in one form or another. And with a further promise of new development, humans can achieve even greater heights than we can only imagine today. Other experiments carried out later to test some of the weird and crazy quantum physics concepts only served to further prove Plank's concept. Presently, the concept of quanta has been applied in understanding space and time and is so far believed to be true. Therefore, you could say that space and time are quantized. However, the current level of technology is not capable of testing for the smallest unit of space and time. We are still decades or maybe centuries away from getting to that level.

Despite how taken the scientific world is today with Plank's model, it wasn't so in the 1900s. No one believed that the amount of radiation energy emitted or radiated could be measured in quanta. There is no way power should exist in simple, indivisible units. Even five years after Planck suggested his model, everyone believed it to be just a temporary method of understanding radiation that only worked because of a fluke and that with time, an actual concept that truly explained radiation would come. To make things even funnier, Planck also believed that his approach to radiation was a temporary solution.

The future only served to strengthen Planck's model as the correct approach to understanding radiation. The work of Albert Einstein on what used to be one of the most difficult scientific mysteries of all time, the photoelectric effect, was thanks to Planck's theory of quantization. According to Einstein, as objects absorb radiation, this energy is added to the motion of the atoms present in the object. This adds energy and causes the temperature of the object to increase. In the case of radiation, the energy gotten from the absorbed radiation is also added to each electron present in the object's atoms. As the energy of the electrons rises, they can overcome the force of attraction of the nucleus. Electrons can finally exit their atoms by gaining enough energy from the absorbed radiation from an electromagnetic wave like light (photoelectric effect).

A significant amount of a wave with long-wavelength red light (approximately 650–700 nm) cannot give an electron enough energy to escape its nuclear binding forces. On the other hand, a wave of relatively low power, such as violet light (400 nm), or an ultraviolet wave (400–10 nm), can liberate electrons from the binding force of the nucleus. So even if the energy of a UV wave is 50 times lower than red light, it can still liberate electrons.

The issue wasn't the energy of the radiation. According to Einstein, waves like UV radiation are better at liberating electrons simply because, if Planck's model is applied, the power of the individual units of radiation is higher than red light. This is in reference to Planck's quantization. Einstein also concluded that for an electron to escape its atom, instead of absorbing several quanta of radiation simultaneously, they only absorb a single quantum that gives them the energy to escape the binding force of the nucleus immediately. What Einstein suggested was such a huge step in our understanding of quantum physics that it caused an uproar in the world. He went on to win a Nobel Prize for his discovery of the mystery behind the photoelectric effect. Therefore, while Planck was still undecided about his theory, Einstein recognized its importance and applied it in proposing his law of the photoelectric effect.

Going with Einstein's law, we can summarize that each quantum of red-light radiation is not capable of causing the photoelectric effect. In contrast, the quantum of UV radiation has enough energy to cause electrons to escape their nuclear binding force. The higher the quantum of the UV radiation, the higher the speed of the electron when it leaves the atom because of the leftover energy from liberating the electron. Einstein's theory was a perfect fit for what happens in reality.

CHAPTER 3: QUANTUM MECHANICS POSTULATES—THE RULES THAT CHANGED EVERYTHING!

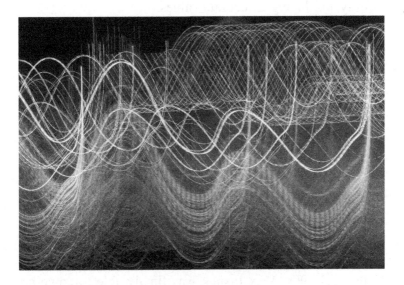

The Principle of Quantization
(Planck, Einstein, and Bohr, 1900)

As we said, black-body radiation simply means the way an object emits electromagnetic waves because of heat. A black body is an object capable of absorbing and releasing all the available frequencies or colors of light.

A black body changes color when heated and releases different frequencies or colors of light as its temperature rises. Classical physics has a law known as the Rayleigh-Jeans law. Under this law, scientists can show how energy changes to heat. However, the law can only be useful at low frequencies. The law isn't useful at higher frequencies closer to the violent light spectrum because it spirals into infinity, producing inaccurate values.

For this reason, it was called the ultraviolet catastrophe. Classical physics also has another law called Wien's law, the opposite of Rayleigh-Jeans law. This law can describe the relation of energy accurately at higher frequencies; however, it produces inaccurate values at lower frequencies.

This issue is called the infrared catastrophe. Both issues of inaccurate values created the need for scientists to develop rules that could help understand this phenomenon.

In the early 90s, Max Planck was finally able to come up with a means to explain this phenomenon. He suggested that rather than looking at a continuum emission spectrum, the electromagnetic radiation was released in little packets of energy. With this, he could introduce the idea of quantization. According to him, specific pockets or quanta of energy are released at a time. This theory of how energy was radiated in quantum is summarized in Planck's constant. And the value is $6.62607004 \times 10^{-3}$ ^{4} J/s (J/s = Joules per second, and Joules is a unit of energy).

The value may be a little scary but applying it is easy. To calculate the energy of the photons emitted by a black body, you simply need to multiply Planck's constant with the photon frequency in question. This can be described by the equation below:

$E = hv$

Where E = Energy of the photon

h = Planck's Constant

v = The photon's frequency

One of the purposes of Planck's constant is to define the energy chunk between frequencies by describing the threshold the frequency of light needs to reach to radiate more energy from a black body. Planck didn't believe much in his theory, as we have described earlier. Albert Einstein and other brilliant minds took up his assumption and found it to be correct.

The Principle of Duality (De Broglie, 1924)

While many great discoveries were happening in the 1900s, scientists began to wonder if all matter could behave like waves. The scientist that worked to solve this puzzle is De Broglie. He came up with a hypothesis that supported the idea that all matter can behave like waves. His hypothesis also showed a connection between the momentum and the wavelength of matter. After Einstein came up with his theory of photons, other physicists tried to apply his theory to all matter to determine if they could display wave-like properties.

De Broglie's Thesis

Louis De Broglie considered the link between the momentum and wavelength of matter when coming up with his hypothesis. According to him, the wavelength of all matter can be determined by the relationship between momentum and wavelength. The formula he came up with is:

$\lambda = p/h$

Where λ is the wavelength associated with the particle of momentum p and h is Planck's constant.

He based his hypothesis on momentum instead of energy because he assumed that the energy of the matter isn't always clear. The energy could be anything ranging from total relativistic energy to kinetic energy. By basing his equation on momentum, he created a relationship with kinetic energy for frequency.

Ek: $F = Ek/HA$

Electrons

In 1927, Clinton Davisson and Lester Germer conducted experiments at Bell Labs to determine if De Broglie's hypothesis was right. During the experiment, they fired slow-moving electrons at a crystalline nickel target. While both of them were doing their experiment, George Paget Thomson, a scientist at the University of Aberdeen, was also doing his own research to also find proof for De Broglie's hypothesis.

The result obtained from both experiments showed that De Broglie was indeed correct. Both studies showed that matter could also undergo diffraction, a property scientists thought only waves could exhibit.

The result of both investigations became the foundation for quantum mechanics. Many more experiments were carried out later on, and the results from all of them only proved that De Broglie's theory was correct. In just the same way the photoelectric effect showed that light could behave as both a wave and a particle, Davisson and Germer proved that matter can also behave as a wave and particle. Both studies also added more proof to the wave-particle duality. For that reason, scientists concluded that particles or matter also display wave-like properties and that this phenomenon can be explained by applying the De Broglie wavelength.

De Broglie was given the Nobel Prize for his revolutionary theory in 1929. In 1937, Davisson and Germer were given a joint Nobel Prize for their pivotal experiment that proved that electrons could undergo diffraction (something De Broglie predicted in his theory). Several studies on diffraction also added to the many experimental proofs of De Broglie's hypothesis.

The Significance of the De Broglie Hypothesis

The most important significance of the hypothesis was that light isn't the only thing that can behave as both a wave and particle. Wave-particle duality was a fundamental property exhibited by both matter and radiation. This means that the behavior of a material can be explained with equations used for waves, as long as you can apply De Broglie's wavelength accurately. This theory became the spring for further discoveries in quantum physics.

Erwin Schrödinger:
The Father of Wave Mechanics

Erwin Schrödinger is one of the most popular names in quantum physics. You probably have already heard of him since he's more popular for his thought experiments than for his scientific experiments. The Australian-born scientist moved to Ireland, where he completed his career even though he had a little run-in with the Nazis. Unlike many other scientists that were interested in quantum physics, Schrödinger's refused to embrace it early in his career. When he did, he created various controversies, which led to the famous feud with Albert Einstein. We will look into his famous cats later but first, let's talk about his work in the laboratory throughout his career.

Schrödinger's Quantum Theory and Unified Field Theory

Despite knowing about the work of Plank, Einstein, and many others, Schrödinger still stuck to his belief that classical physics could still produce results. However, after he fought a tuberculosis infection, his stance on quantum physics changed. While he was recovering, he started becoming interested in quantum theory. He became obsessed with the possibility that the electromagnetic spectrum could affect numerous elements in several different ways. He got interested in wanting to find out if he could discover a universal method to determine the behavior of electrons. While he was thinking about this, he finally summarized that the only way he could achieve what he wanted was to accurately predict the nature of radiation. Although he wasn't able to predict the nature of radiation himself, his work in trying to predict the nature of radiation was important in laying the groundwork for a new type of theoretical physics. This new brand of physics includes wave mechanics and another atomic model.

Schrödinger also contributed immensely to creating a unified theory which is considered the Holy Grail of quantum physics. If he had succeeded in creating a unified field theory, he would have unified all existing relationships within quantum physics under a single set of laws. This will allow researchers to accurately predict the behavior of every matter using mathematical proof and observation of action without any deviation.

This law would be able to predict the behavior of matter, and scientists would finally be able to explain the relationship of electromagnetic fields, particles, and space-time with one another. Despite many attempts to develop a unified field theory for quantum physics the same as classical physics, no one has been able to do it. One of the most popular theories to attempt the feat is known as the "theory of everything," proposed by Stephen Hawking in 1971 which focuses on a series of basic knowledge about the black hole mechanics. According to him "the total area of a black hole's event horizon—including all black holes in our universe, for that matter—should never reduce." This was explained in his book titled *The Theory of Everything*.

Wave Mechanics

Schrödinger succeeded in becoming the founder of wave mechanics. This led to the birth of a new branch of quantum physics. In the 1920s, he created theories to explain the behavior of particle waves based on how hydrogen atoms would behave in a system free of the effects of time. His goal was to find out the result if time isn't a factor in the equation when you want to predict the atom's behavior as a wave-particle. He released a series of four papers where he wrote down what he predicted by introducing his infamous equation. In his second paper, he edited his equation, taking note of harmonics in the system.

The goal of his third and fourth papers was to show how he can compare the work done with the uncertainty principle to his equation and to show his colleagues how they can correct the complex numbers in his equation to avoid calculating numerous derivatives.

Today, his series of papers is regarded as one of the biggest accomplishments in modern physics and science as a whole. It opened the way for researchers to study the behavior of waves in a more price-controlled way. His papers also marked the release of some complex math with the introduction of his equation which many believed is the point of separation between classical and quantum physics. Now you see why it's hard to pick out a specific defining moment in the history of quantum physics. Despite his work, Schrödinger was over the moon with his achievements after he published his papers. He understood that he had drawn a huge line between classical and quantum physics, especially considering he loved sticking to the principles of classical physics. But there was no going back after he published his papers.

Schrödinger's Equation and the Atomic Model

You have probably seen Schrödinger's equation by now, and you are probably wondering why it is so special. Is it the funny-looking numbers? And why did it have the influence it did on quantum physics?

The first thing you have to know about the equation is a partial linear differential equation. This simply means that you have to consider many moving parts if you want to arrive at the answer. Even the description of his equation is complex, so we will try to explain it to you.

In classical physics, Newton's second law of motion is described by the formula F=ma, which translates to force equals the product of the multiplication of mass and acceleration. This equation describes the movement of an object based on its speed and mass. Schrödinger's equation can be seen as the quantum physics equivalent of Newton's second law. This was because it had the same influence as Newton's second law.

Below is the equation:

$$H(t)|\psi(t)\rangle = i \cdot \frac{\partial}{\partial t}|\psi(t)\rangle$$

Yeah, it's a lot to take in, but don't worry, we are not going into specifics about the equation because this is only a book for beginners. What Schrödinger did was to take F=ma and turn it upside down. You need to look at the equation and work out the math if you are going into quantum physics for academic purposes. If the Schrödinger equation didn't exist, scientists might still struggle to predict how wave functions behave. The equation allows scientists to get the answer they need by doing simple math, and it opened the door to quantum mechanics even though he was a strong believer in classical physics.

Schrödinger also went further to update the then-current atomic model, taking into account electrons' behavior as waves go around the nucleus. He spent time doing analysis and came up with the first true three-dimensional and correct atomic model by applying his theory and equation of wave mechanics.

His atomic model consists of an electron cloud, orbiting inwards and outwards the nucleus in a wave-like manner. With this atomic model, scientists can accurately tell the location of electrons at any time. This was different from the model proposed by Bohr, whose model placed the electrons in set orbits of layers instead of fluctuating following wave-particle duality.

Schrödinger's Cat

Despite his stubbornness to stick to classical physics earlier in his career, not only did Schrödinger come to accept quantum physics, he became probably the weirdest among our long list of quantum physicists.

Apart from his numerous achievements and Noble Prize, Schrödinger is also famous for his untraditional lifestyle: his marriage was a weird arrangement, and his decision to immigrate to Ireland with his wife and mistress alongside the children he had with both women was not normal. The premise he wrote on genetics marked the beginning of numerous inquiries that resulted in discovering the human genome. He also wrote a treatise on the visible light spectrum, which focuses on the nature of colors and how humans perceive colors.

His works showed his curiosity for both science and philosophy. And this was apparent in his thought experiment, Schrödinger's Cat, for which he is mostly known today.

Schrödinger's cat theory proposes a strange situation in quantum physics: Imagine a cat in a box with a device that can kill it with a deadly poison. The theory posits that, as long as the box is not opened to check whether the cat is alive or dead, the cat would be both alive and dead at the same time, in a kind of quantum state. In this way, it created a paradox. With the box closed, it is impossible to know whether the cat is alive or dead without opening the box to look at it. He also thought that opening the box might interfere with the intended outcome.

This concept is based on the principle of quantum superposition, which indicates that a subatomic particle can be in several states at once until it is measured or observed. Schrödinger's cat theory is a way of illustrating how this principle applies to macroscopic objects, such as a cat, which in the theory is considered a quantum system.

Schrödinger introduced his cat through an experiment during his debates with Einstein. He created the thought experiment as a gentle way of fighting against the way certain scientists interpreted his theories on quantum mechanics. He was angry against what is known as Copenhagen's view of his theories.

The Copenhagen view of his theories headed by Bohr said that Schrödinger's wave function would collapse because it is impossible to view the nature of waves without it leading to terrible results due to interference from the observation. Bohr, alongside several other scientists, suggested that attempting to measure would interfere with the experiment, which would make the result meaningless. In his defense, Schrödinger argued that Bohr couldn't prove his argument because no one has physically observed a collapse. Out of frustration, he finally came up with his thought experiment with the cat. His goal was to show his colleagues that their interpretation was wrong. Although Schrödinger agreed that the observer effect does exist, the simple act of looking at something is not enough to affect it drastically. To explain his side of the argument, he placed a cat in a box and asked his colleagues to guess if it was dead or alive. Even though many people saw his display as outrageous, it had a ripple effect in the science world and on quantum physics.

Although Schrödinger finally gave in that maybe his theory cannot be right all the time, he concluded that what Bohr and his friends were trying to explain wasn't a collapse but rather wave entanglement. In the same way that kids play cat's cradle and get the yarn knotted, the waves of electrons can also get entangled as they orbit independently around the nucleus. The introduction of quantum entanglement opened a new door to wave mechanics, which allowed scientists to know what they should find as they studied the behavior of subatomic particles.

In 1961, Erwin Schrödinger lost his fight with tuberculosis and was buried in Austria. His famous equation was engraved on his tomb, where it remains. His wife died in 1965 and was buried next to him. As for the mistress, no one knows what became of her.

Schrödinger will always be remembered as a man that followed his heart. He only took up quantum physics when it was OK for him, contributing immensely to its development, and left to chase other dreams when he wanted. He didn't care about the invasion of his homeland by Nazis and informed them of his stance. Everyone knew about his open marriage, and he didn't care, and he created a way to talk about cats being dead in a scientific way. During one of his last public meetings, he was tasked to talk about nuclear power. However, he declined and talked about philosophy instead until there were protests from his audience. The incredibly talented and brilliant scientist will always be remembered not just for his work in quantum physics but also for his cats.

The Principle of Uncertainty (Heisenberg, 1927)

One major difference between classical and quantum physics is their theory of the measurement process. You can know the position and momentum of a physical system by simple and precise measurements. In classical physics, the system is not affected by measurement, and anyone can measure infinite weight with precision. However, this is different for quantum physics because the measurement affects the system.

When you measure every infinite copy of this system simultaneously, you will get the probability distribution of all values you can measure. The probability of every value is equal to the square of the coefficient of the correlating eigenstate which is the measured state of an object that possesses a measurable or quantifiable property like momentum or position.

The most popular observables that are incompatible include the momentum (P) and its position (x). When you multiply both observables, that is Δx and Δp, the result you get is either greater than or equal to half of Planck's constant.

Heisenberg came up with the constant uncertainty principle in 1927. In a physical relationship that has two mechanical quantities such as in momentum and coordinates, which can be represented by two misaligned quantities, you can't get two absolute measured values of these quantities simultaneously. As one of the values becomes more accurate, the other becomes less accurate. This means that because the measurement process can interrupt the behavior of microscopic particles, you can't interchange the sequence of measurement. This is a primary law of microscopic phenomena.

This goes on to show that the information of physical quantities like the momentum and coordinates of particles don't already exist and are to be measured. Measurement is a transformational process rather than a process of reflection. The value we get in measuring such physical quantities depends on the method used. Therefore, the uncertainty relationship is the result of the mutual exclusion of methods of measurement.

We can get the potential amplitude (the number used to describe the system's behavior) at each eigenstate by breaking down a state into a linear mixture of eigenstates we can observe. The probability of a measured eigenvalue is the fixed square of the possible amplitude, which is the probability that the system is in the eigenstate. In conclusion, this means that we can get different results from the same measurement of a quantity we can observe in a similar system.

CHAPTER 4: QUANTUM ENTANGLEMENT

In his many fights with Einstein, Schrodinger came up with the concept of quantum entanglement, which he introduced in 1935 and 1936. He presented this concept in a two-part paper in proceedings to the Cambridge Philosophical Society. The goal of the paper was to explain and expand the theories of Podolski, Rosen, and Einstein.

Einstein and a host of other scientists criticized Schrodinger's equation, and their goal was to prove that his equation could lead to collapse and therefore was incomplete. Even though we have made rapid advancements in science, we still don't know numerous details about the universe. Our curiosity has only created more mystery and points we lie in bed wondering about. The day that we begin seeing ourselves as the ultimate supreme being is the day we begin to run ourselves. This is one of the major driving forces that led to the discovery of the concept of quantum entanglement.

Many people, including scientists, consider the entanglement theory a special phenomenon, but the truth is its fundamental issues are easy to understand. The best way to understand the entanglement theory is to think about it in terms of classical physics first. Although going by this route is unconventional, it enables us to understand it faster.

INDEPENDENT

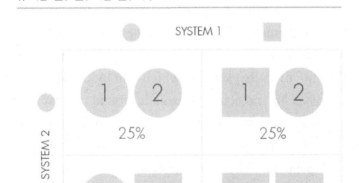

The T-ins are independent if data about one of the states don't provide any useful knowledge about the states of others. The table above can be referred to as independent.

The T-ins can become entangled when our knowledge of one gives us information about the state of the other.

The quantum explanation of entanglement happens in the same way; in other words, there is a lack of independence. Mathematical objects known as wave functions are used to represent states in quantum theory. It is more complicated by the rules that link wave functions to physical probabilities, as we will see below, but our classical understanding of entanglement still applies. Quantum systems usually experience entanglement naturally. For instance, in the event of a collision of particles. In reality, it is rare to find independent states because the interaction between systems leads to a linkage between them. We can look at an example of this in molecules. A molecule is made up of subsystems, which include nuclei and electrons. The molecule is usually found in its lowest energy state, which is an entangled state of both its nuclei and electrons because the positions of both particles are not independent. When the nuclei move, then the electron follows.

Let's go back to our example: say we represent one system with the wave function, $\Phi\blacksquare$, $\Phi\bullet$ which represent its circular and square states, and another system with the wave function $\psi\blacksquare$, $\psi\bullet$ showing its circular and square states. This means that the total state we will end up with will include:

Independent: $\Phi\blacksquare\ \psi\blacksquare + \Phi\blacksquare\ \psi\bullet + \Phi\bullet\ \psi\blacksquare + \Phi\bullet\ \psi\bullet$

Entangled: $\Phi\blacksquare\ \psi\blacksquare + \Phi\bullet\ \psi\bullet$

We can also represent the independent version with:

$(\Phi\blacksquare + \Phi\bullet)\ (\psi\blacksquare + \psi\bullet)$

We can create entangled states in many ways. An example is to measure a composite system to get some information. One piece of information we can learn is that the two systems have chosen to have similar shapes without knowing what shape is in particular. There are several consequences of quantum entanglement, which include Greenberger-Horne-Zeilinger (GHZ) and Einstein-Podolsky-Rosen (EPR) effects. These effects are the result of the relationship with another part of quantum theory known as "complementarity." Before we look into GHZ and EPR, let's talk about complementarity.

Instead of imagining that our T-ins have two shapes, think of them as having two different colors, blue and red. If we go back to our classical understanding, then there are four potential states: a blue square, a red circle, a red square, or a blue circle. In quantum entanglement this is different. Because a T-ins is capable of having different colors or shapes depending on the situation doesn't mean that it can have both a color and shape at the same time. Although

Einstein insisted that this should be accepted, results from experiments don't support it.

When we measure the shape of T-ins or objects, we lose data about their color in doing so, and the same can be said when we measure their color. According to quantum theory, it is impossible to measure both its shape and color at the same time. We cannot capture all the aspects of physical reality from one view. To get the full picture, you need to consider all the mutually exclusive events possible —this formed the basis of Niels Bohr's principle of complementarity.

By following quantum theory, we cannot just assign physical reality to each property. To prevent contradiction, the following rules must be obeyed:

- If a property is not measured, then it doesn't exist.

- The process of measurement changes the system that is being measured.

One of the most popular effects of quantum entanglement was described by Albert Einstein, Boris Podolsky, and Nathan Rosen (EPR). This effect described the relationship of quantum entanglement with complementarity.

An EPR pair refers to two T-ins, both of which you can only measure their color or shape but not for both. Imagine that you can assess numerous similar pairs and you are completely free to measure any of their components. This means that when you measure the shape of a pair, it can either be circular or square. And when you measure for the color, it can either be blue or red.

The paradoxical effects found by EPR occur when you measure both parties in a pair. When both the colors and the shape of both members of a pair are measured, the results always go together. So, if the color of one member in a pair is red, the color of the other member in the pair will also be red. This phenomenon also happens in the case of shapes too. On the other hand, when you measure the color of one party and the shape of the other, they don't go together. If the shape of the first member of the pair is square, the color of the second pair will probably be blue or red. According to quantum theory, distances between both members have no effect on the two systems if the measurement is done simultaneously. This phenomenon seems to be a result of the method of measurement affecting the state of the system in another location. Einstein named this effect "spooky action at a distance." He also concluded that it occurs as a result of the transmission of information about what measurement is done at a rate that is faster than the speed of light.

The paradox goes deeper with a deeper reflection. If you measure the state of the first system and get red, then you will also get red when you measure the color of the second T-ins. But when you introduce complementarity, when you measure the shape of the T-ins when the color is red, there is a 50–50% chance that you will have a circle or a square. This way, you can see that the effects as dictated by EPR are logically forced.

Another effect we are going to look into is the one introduced by Daniel Greenberger, Michael Horne, and Anton Zeilinger. This effect involves three T-ins in a unique state known as the GHZ state. The three T-ins are distributed in three separate experiments. Each experiment is free to randomly choose to measure either the color or the shape. The experiment is repeated as many times as possible. After performing the experiment many times, the results from all three procedures are compared. The results revealed an outstanding peculiarity. We are going to name red colors and all squares "good," then blue colors and all circles "bad." The result showed that precisely zero or two were either blue or circular (bad) each time two experiments measured for shape and the remaining experiment measured for color.

However, when all three measurements for color, precisely three or one were squares of red (good). This prediction is also observed in seen mechanics. The question now is what is the quantity for bad: odd or even? What you get depends on the method of your measurements. Therefore, the quantity of bad in our system depends on how it is measured.

The term annihilation refers to the process of conversion of a particle and its related antiparticle to electromagnetic radiation (photon) or other particles. In the case of pair creation, a particle and its antiparticle are created by the interaction of electromagnetic fields or any other field. This is seen in the collision of an electron and its corresponding antiparticle (positron), which leads to the generation of two photons.

When a proton collides with an antiproton, the result is inter-annihilation accompanied by the creation of other lighter particles and pi-mesons (pion). The function of a pi-meson is to hold the nucleus together, which is why they are produced when there is a high-energy collision between particles. A gamma quantum with enough energy can produce an electron-positron pair by interacting with the electric field of a nucleus. In conclusion, the creation and annihilation of particles is all about the interconversion of particles rather than the abrupt evolution of matter. The fundamental laws of conversion include the laws of conservation of energy and momentum, electrical charge, and angular momentum. The English physicist P. Dirac was the first person to consider the idea of pair annihilation and creation, including the existence of antiparticles.

The Annihilation of an Electron-Positron Pair

A positron will lose almost all its velocity because of a loss of energy when it collides with matter. The positron is at rest when it undergoes annihilation. An electron or positron programmed for destruction is in a state with an angular momentum of zero. The fate of the particles depends on both their orientation of the angular momentum (their spins)

When the spins of the positron and the electron are in opposite directions, it will lead to the creation of an even number of protons because of annihilation. It is impossible to create an odd number of photons because of the law of conservation of energy. There is still the likelihood of creating more than four photons, although it is small.

The created photons jump in opposite directions and each of them receives half the initial energy of the system of electrons and positrons. Following A. Einstein's theory of relativity, a particle at rest with a mass M has an energy E0 = Mc2, which is known as its rest energy.

The Creation of an Electron-Positron Pair

To create an electron-positron pair, with the aid of photons, you need an extra or external electromagnetic field that will function as a second photon. This follows the laws of conservation of momentum and energy. This simply means that it is impossible to transform a single photon into a particle-antiparticle pair. This reaction that forms the particle-antiparticle pair usually occurs in the Coulomb field of an electron (or an atomic nucleus). For this reaction to happen, the energy of the photon must be above the sum of the rest masses of the positron and electron. This is given mathematically as: $2mc2 = 1.02$ MeV.

The likelihood of creating a particle-antiparticle pair in a coulomb field of a nucleus is directly proportional to the square of the charge of the nucleus, $Z2$. This probability increases exponentially as the gamma-quantum energy increases and gets to a finite value at extremely high energies. The creation of a pair is important in the absorption of gamma quanta of high energy by matter.

The Annihilation and Creation of Other Particle Pairs

When a photon has very high energy, it can create other types of particle-antiparticle pairs. A pair of particles interacting strongly (e.g., a proton-antiproton pair) can be caused by extremely fast protons colliding with nucleons of a nucleus. During the annihilation of nucleons and antinucleons, there is a more constant evolution of gamma-quanta as well as massive particles. The law of conservation does not prevent this occurrence. Four or five pi-mesons are always formed in the annihilation of such pairs.

CHAPTER 5: QUANTUM TUNNELING

In one of the previous chapters of this book, we have already covered that all matter can behave as a wave. And we did this by talking about a study that proved that particles in a matter, particularly electrons, can undergo diffraction.

The phenomenon of quantum tunneling happens when two particles can go through a barrier, which is supposed to be impossible in the classic physics realm. The barrier can be a physically solid impossible medium like a vacuum or insulator or a region with high potential energy. Particles with insufficient energy can go through a potential barrier in classical physics. However, quantum physics particles behave like waves, making it possible to go through a barrier. When quantum waves contact a barrier, it doesn't just end abruptly. Instead, its amplitude reduces exponentially. The resultant decrease in amplitude increases the difficulty of finding the particle as you move further in the barrier. If the barrier is thin, the amplitude value may be greater than zero on the other side. This also means that there is a fixed probability that certain particles will tunnel through the barrier.

The definition of quantum tunneling is the ratio of the density of a wave that makes it out of the barrier divided by the density of the incident wave on the barrier. If the transmission coefficient through the barrier is above zero, then the probability of a particle tunneling through the barrier is finite.

The Discovery of Quantum Tunneling

The first person to consider the existence of quantum tunneling was F. Hund in 1927 while he was performing calculations on the ground state energy in a system called double potential. Two similar energies with two different states are divided by a potential barrier in a double potential system. An example of this type of system is the existence of ammonia molecules.

Ammonia molecules often take on the shape of an umbrella, where three atoms of hydrogen fan out in a nonplanar placement with a centrally placed atom of nitrogen. Because of the stability of this umbrella structure, a large amount of energy is needed for it to be inverted. But quantum tunneling has made it possible for ammonia molecules to possess geometric structures separated by a barrier of high energy.

There is no such thing as the "inversion" transition between two separate geometric states in classical physics, but it exists in quantum physics.

L. Nordheim also observed another event involving the tunneling phenomenon around the same year as F. Hund. He discovered one of his studies on electron reflection from various surfaces. Oppenheimer became the first to use the tunneling phenomenon to calculate the hydrogen ionization rate within the next few years. Garnow, Gurney, and Condon did other prominent studies on the phenomenon. Three of them explained the extent of the alpha decay rates of radioactive nuclei.

Quantum Tunneling in Nature and the Sun Fusion

Many people seem to think that quantum physics has no bearing on our everyday lives, which also applies to quantum tunneling. However, quantum tunneling is an important part of nature that plays a huge role in our daily lives. According to several theories and hypotheses, the actual start or origin of the universe is from a tunneling incident. This allowed the universe to move from a "state of no geometry" (space and time) into a state that allows the existence of space, time, and matter which is necessary for life. Quantum physics generated the classic natural physics laws that we are used to live by.

Tunneling in the Sun and Other Stars

You must have learned about nuclear physics in high school, and if you paid a little attention, you could remember hearing about nuclear fusion. As the name implies, fusion simply means combining two or more things. In nuclei, fusion is the combination of small nuclei to make an even larger nucleus, which causes the release of large amounts of energy. The type of fusion that occurs in stars leads to the production of all the elements we have in the periodic table, except for energy. Stars get their fuel from the fusion of hydrogen, which produces helium.

Scientists initially thought that fusion was a rare occurrence, but today the world is aware that it occurs more commonly than we thought. As you can expect, nuclei repel one another strongly because they are positively charged. And they usually don't have enough kinetic energy to overcome the repellent force, which will lead to fusion. However, when one considers the effect of tunneling, the percentage of hydrogen nuclei that goes through fusion increases massively. The link between tunneling and nuclear fusion explained the reason stars can exist in stable states for thousands to millions of years. Science does not fully support this process because an average hydrogen nucleus has to go through more than 1,000 head-on collisions before finally fusing with another nucleus.

You Use Quantum Physics to Smell!

Scientists used to think that nose receptors detect chemicals by the lock and key mechanism, meaning that the brain can tell the different types of chemical smells by the shape of their molecules that binds to the corresponding nose receptor. In other words, the 400 different types of receptors present in the human nose can tell various chemicals by identifying their shape as they bind to the receptor with their corresponding shape similar to how a key opens a lock with its corresponding shape. However, today scientists have discovered several problems with this previous belief. For instance, ethanol and ethanethiol molecules have closely related shapes, yet they smell very differently. Ethanol is the major component of alcoholic drinks, so it smells like alcohol, while ethanethiol has the sulfuric smell of rotten eggs. This led scientists to think that the mechanism involved in smelling is not what they thought.

The recent theory uses quantum tunneling to explain how receptors detect and relay chemicals. According to this theory, the receptors release a small current to the odorant molecule, which makes it vibrate uniquely. For the current to flow from the receptors to the molecule, electrons must turn into the gap between the cells of the molecule and the receptor, which is incapable of conducting electricity. Recent experiments to prove that quantum tunneling plays a role in smelling involve the use of large deuterium and hydrogen to increase the reaction of the olfactory to stimuli. The experiment's result showed an increase in quantum vibrations, meaning that humans can perceive different smell molecules using their unique quantum vibrations. Each smell molecule has a unique vibrational pattern which the nose receptors can differentiate therefore enabling the brain to tell them apart. Today it is thought that the two theories work together to give us the full experience of smelling.

Applications of Quantum Tunneling

Josephson Junctions
A Josephson junction refers to a junction formed by two superconductors separated by a super thin layer of a non-superconductor. This non-superconductor can be a physical defect, an insulator, or a non-superconductor. The current in the system can move across the superconducting materials. This makes the system's electrical properties to be accurately precise.

This type of application has opened the doorway to various special constructions, majorly for making accurate measurements. Several systems where you can find Josephson junctions include quantum computers and other superconducting devices. It is also applied massively in superconducting quantum interference devices (SQUIDs). These devices can measure extremely weak magnetic fields. There are also several studies on applying these junctions to measure quantum coherence, a phenomenon that occurs in the world of subatomic particles, such as atoms or photons. In these systems, particles can be in multiple states at once and these states can interact coherently.

An example of quantum coherence is when light passes through a slit and forms an interference pattern. This is because the photons of light pass through both slits at the same time and, because of quantum coherence, they can interfere constructively or destructively.

Tunnel Diodes

It is made of two semiconductors separated by a thin insulator. The tunnel diode is also named after L. Esaki, and it is called the Esaki diode as a result of his achievement in making the diode what it is today.

A diode is an electronic device that allows electric current to pass in one direction and blocks it in the opposite direction. It is a basic component of electronics and is used in a wide range of applications, such as current rectification, reverse polarity protection, signal modulation, and light generation.

The diode can generate and work with frequencies in the microwave spectrum. This is a piece of discovery that has made the construction of many supercomputers possible because it allows the processing of data. We will see how it makes this possible and how it applies quantum tunneling.

It is a device with two terminals, and it has a massively doped p-n junction where electric current flowing through the device occurs due to quantum tunneling. The current flowing across the diode decreases significantly with increasing voltage. The diode responds to this by exhibiting a negative resistance. The impurities in the tunnel diode are 1,000 times more than that in a standard p-n junction diode. Therefore, the p-n junction exhibits a significant narrow depletion region of only nanometers. The current method in standard diodes is that the voltage applied is more than that of the deletion region. However, in tunnel diodes, when a small voltage less than that in the deleted area is applied, it generates an electric current due to quantum tunneling between the regions of p and n. The narrow depletion region is that the thickness of the barrier is just enough for tunneling to occur. Devices that use quantum tunneling include relaxation oscillator circuits, ultra-fast switches, and logic memory storage devices. They are also frequently used in the nuclear industry because they are highly resistant to radiation.

Scanning Tunneling Microscopes

The Scanning Tunneling Microscope (STM) is another device that works by applying quantum tunneling. This device operates by scanning across the surface of a material for a sharp conducting probe. An electric current is made to flow downwards the probe's tip before tunneling through the gap into the material. The current can increase or reduce depending on how narrow or wider the gap gets. Scientists can build a well-detailed surface, even free of humps due to atoms, thanks to this data. This has contributed massively to our understanding of surfaces.

Flash Drives

Have you ever wondered how just a little device like your flash drive records and stores your data? Have you ever thought that maybe quantum physics plays a role in the mechanism involved in storing data? The ability of flash drives to store data is due to their memory cells made of "floating-gate" transistors. These transistors contain two metal gates: floating and control gates. An insulating layer made of metal oxide surrounds the floating gate.

In its usual state, the floating gate records a "1" in binary code. An electron is attached to a floating gate, which traps it in an oxide layer. This changes the voltage in the control gate. In this state, the transistor records a 0 in binary. When you erase the data from your flash drive, a powerful positive charge is made to flow through the control gate. The trapped electron in the control gate responds to this positive charge by tunneling through the insulating layer.

CHAPTER 6:
QUANTUM COMPUTING

In today's world of computers, it seems that we have found a way of getting more and more into smaller and smaller sizes. It seems that as computers get smaller, they become more powerful. Your current cell phone can perform better than cell phones in the 20th century.

The same could also be said for those big computers used in the 20th century. Despite these rapid advances in computers, we are still far from solving complex problems even with the world's most powerful supercomputers. Scientists need to go out of the usual method of operating if they want to achieve their goal of computers that are more powerful and smaller than the ones we have today.

Our discovery of the world of atoms opens the door to several possibilities in the frame of quantum computing capable of processing data thousands to millions of times quicker than the ones we have today. Yeah, this sounds amazing, but it has its issues because a quantum computer is massively more complicated than conventional computers. This is because quantum computing is almost impossible without a perfect understanding of what quantum physics is all about, and we have yet to achieve this fully. The usual physics or classical mechanics we know don't apply to quantum physics, so we have to think outside the box for us to achieve quantum computing.

What Is Conventional Computing?

Your imagination of quantum computers might be that of a cute little gadget you can use while lying on your bed to send emails, talk with friends, shop online, or play video games. Although we can do all this when we finally have a quantum computer that everyone can use in the world, it is more complex than we can imagine. Quantum computer scientists are building a small do-it-all machine. Although you can do anything you like with it, it will still be able to use complex calculations that the supercomputer can do. The quantum computer is a bit more than just a complex calculator that follows a preset program.

The conventional computer you are currently using has two things they are very good at. They can store numbers in memory and process these values using several mathematical operations such as subtraction and addition. They can even perform several complicated operations by joining several simple operations to form a series known as an algorithm. These processes are possible thanks to the existence of transistors. You can think of transistors as microscopic switches for putting lights on and off on the wall. These transistors can exist in two states: on or off. Transistors install a number one in the "on" state (1), while the "off" state stores the number zero (0), and then they use long strings of zeros and ones to store letters, symbols, or numbers. The zeros and ones are known as binary digits or bits, and these strings of eight bits are used to record 255 different characters, including 0-9, A-Z, and various symbols. To perform calculations, computers use circuits called logic gates which consist of several linked transistors. A logic gate operates by comparing various patterns of beats recorded in temporal memories known as registers which turn into new bits patterns. This computer operating system is the equivalent of doing additions, subtractions, and multiplications in the human brain. Physically, the algorithm responsible for doing a particular calculation is a group of logic gates that forms an electronic circuit. One gate functions as the output that feeds in the input to the next one.

The major issue with today's computers is that they work using conventional transistors. However, this problem might not seem like much of an issue considering how much impact humans have made using electronics in the last few years. Transistors were first discovered in 1947, replacing the switch used, then known as a vacuum tube, as big as a human adult thumb. Today, the world's best computer chip or microprocessor comes with hundreds of millions to even billions of transistors on a silicon chip the size of your nail. These chips are known as integrated circuits, and they are a marvel of modern technology. In the 1960s, the co-founder of Intel, Gordon Moore, discovered that the power of computers increased by two approximately every 18 months, and this has been the same since then. This phenomenon is called Moore's Law.

Despite how amazing this sounds, it still has limitations because you require even more binary zeros and ones if you want to store more data, which translates into the need for more transistors. Many steps are involved to accomplish very complex tasks, and today's traditional computers may take a longer time to process them. There are even more complicated operations that need more computing time and power than any present-day machine can provide. These kinds of issues are called intractable problems.

The number of intractable problems reduces more than Moore's Law advances. Therefore, computers are expected to become more powerful to solve more complex tasks. However, science has hit a wall at the moment because we can't make transistors smaller than they currently are, which means Moore's Law is coming to an end. We still have many problems to decipher yet because even the power of supercomputers is not enough to solve them. For this reason, we are turning to quantum computing.

What Is Quantum Computing?

Going back to Moore's Law, we would have to make smaller and smaller versions of transistors until they reach the point where conventional laws of physics no longer apply to them. Many people wonder if we can make our computers achieve the impossible. Trying to guess the future of computer achievement, the big question is: Will we be able to bring to life a new generation of computers substantially different from conventional ones?

This question has plagued many scientists for decades, but the first ones to ask it were Charles Bennett and Rolf Landauer. Landauer got the world thinking about quantum computing when he suggested that information is nothing but a physical entity that we can control using the laws of physics. We see this in how conventional computers waste energy controlling the bits within them, which is also responsible for the massive energy used by computers (as a consequence, computers can get hot even though they don't seem to be doing much work). Bennett followed the idea suggested by Landauer and showed the world how a computer can solve this issue by operating reversibly. He showed in the 1970s that quantum computers can solve even the most complicated tasks with minimal energy consumption and without the energy loss of conventional computers.

Paul Benioff picked up from where Bennett left off in 1980 by trying to suggest a basic device that would behave similarly to a conventional computer but that worked using the actual principles of quantum physics. This device was seen as a quantum Turing machine. The next year, Richard Feynman came up with a rough sketch of the way a machine can apply quantum principles in dealing with simple computations. The next few years after that saw Oxford University's David Deutsch, one of the brightest minds in quantum computing, coming up with the theoretical foundation of how quantum computers could work.

Now, the major unique characteristics of conventional computers are bits, logic gates, and registers; these and other features have analogs in quantum computing. Rather than bits, quantum computers use quantum bits (qubits) that work in a similar way. A bit in conventional computers can store only one of zero or one; qubits can store a one, a zero, both one and zero, and an endless number of values in between. While this may seem confusing, if we think that light exists as both a particle and wave simultaneously or the fact that Schrödinger's cat can both be alive and dead, then it begins to make sense. To make things easier for you, we will look at the concept of superposition (where you can add two waves to get a third one containing both of the original waves). When you blow a musical instrument such as a flute, the pipe is filled with a standing wave. This type of wave has a basic frequency, the basic note you are playing, and several overtones, which are higher-frequency multiples of the original frequency. The wave in the pipe is filled with these waves at the same time. The addition of all the waves results in a wave that contains all of them. By using superposition, qubits can store more than multiple numbers at the same time in the same way.

Similarly, quantum computers store multiple numbers at the same time; they can also process them simultaneously. Unlike conventional computers that process data in series by working on one task at a time and can only process a series at a time, quantum computers can handle several series simultaneously. It does this by processing data in parallel. Quantum computing can collapse into one of its likely states when you try to find out what state it is in at any moment. According to this theory, the ability of quantum computers to operate parallel tasks makes them millions of times quicker than a traditional computer. The problem is that we are still far off in building a complete quantum computer, although companies like IBM already attempted and succeeded to build a non-commercial machine. Recently, IBM also unveiled a new quantum pool chip that contains a milestone of 127 quantum bits, or qubits.

Quantum Dots in Computing and Other Applications

The data in qubits can only be stored in atoms, ions, or smaller particles like photons (energy packets) and electrons. If we can achieve this, then we would need a way to contain these particles that are going to serve as a store of energy. The mechanism of storage for these particles will allow them to move into certain states that allow them to process information and determine their current state after certain operations.

Some experiments have shown that there are ways we can use to store or contain these atoms. There are also ways to change the states of these storage particles with the aid of electromagnetic fields, laser beams, radio waves, and other methods. One popular recent technique used to achieve this is to create qubits with the aid of quantum dots. Quantum dots are nanoscopic particles in semiconductors thanks to which we can control individual charge carriers, holes, and electrons.

Another technique used to make qubits is known as ion traps. This method involves adding and removing electrons from an atom to create an ion, then holding it steady within a laser spotlight, and then changing it to various states using laser pulses.

A more recent technique involves storing the qubits as photons in optical spaces (cavities in very small mirrors). You don't have to understand every method used for storing and changing the states of these particles; after all, very few people know what they are all about. Almost all parts of quantum computing are still theoretical and abstract. You only need to understand that qubits can be stored in atoms or other subatomic particles that can be changed from one state to another to store and process information.

How Quantum State Will Make Humanity Evolve

Many people just expect quantum computers to be so much better than traditional computers, but the reality is, that there is no clear "king of computing." Currently, the only thing that we can be sure of is that quantum computers can perform better when it comes to factorization. This involves looking for two prime numbers, both unknown, which you can multiply to get a third already determined number. Peter Shor discovered an algorithm that a quantum computer could use to search for the prime factors of huge numbers during his work at Bell Laboratories in 1994. Doing this would significantly shorten the solving process of the problem.

Shor's discovery sparked a lot of interest in quantum computing, majorly because almost all modern computers, including secure banking sites, use encryption technology that depends on how impossible it is to quickly find prime factors. If quantum computers exist and are capable of solving factorization problems, then our online security today will be meaningless. But researchers predict that more complex and stronger encryption methods will arise when this happens.

Despite Shor's discovery, it doesn't mean that quantum computers can make our lives so much better. There has almost been no other discovered algorithm that quantum physics can do better inherently than Shor's and Grover's algorithm. All conventional computers need is time and power, and they will be able to deal with any problem that can be solved with quantum computers. In conclusion, no one has proven that quantum computers are entirely better than conventional computers, especially when you consider the difficulty involved in building them. Conventional computers still have a lot of room to grow and could advance to unbelievable levels in the next 50 years, which could mean that there may be no need for quantum computers at all.

Why Is It So Hard to Make a Quantum Computer?

Despite our years of experience manufacturing conventional computers that work using transistors, we can't apply this experience when it comes to making quantum computers. To successfully build a quantum computer, we need to change our entire knowledge for creating a computer. One of the first points we need to learn is how to make qubits and then control them to change states to perform certain tasks. We need to learn how to overcome inherent errors in quantum systems, usually called noise, which can interrupt any calculation we want to do with the quantum computer. Although there are techniques for overcoming these inherent errors, called quantum error correction, this only complicates things.

Another issue is how to put data into a quantum computer and how to get it out. Some critics believe that there is no way humans can overcome these issues, but others believe that we should not abandon the mission as it is too important to disregard.

CHAPTER 7: STRING THEORY, THE TRUE NATURE OF REALITY?

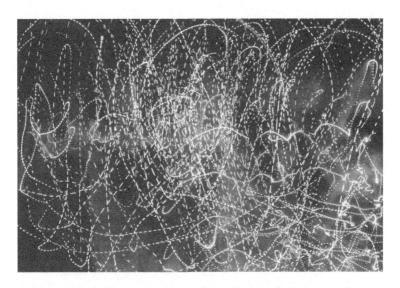

String theory refers to a theoretical structure where one-dimensional objects known as strings replace the point-like particles of particle physics. The theory aims to explain the interaction of these strings with each other and how they are propagated in space.

A string will appear as a simple particle on larger distance scales as opposed to a string scale. The charge and mass, as well as other characteristics of the string on a larger scale, depend on the state of vibration of the string. The theory suggests that one vibrational state of the string is proportional to the graviton, which is a quantum particle that can carry gravitational force. In this way, you can describe string theory as a theory of quantum gravity.

String theory was discovered as a way to provide answers to some mysterious questions concerning fundamental physics. This theory has been used to explain many issues ranging from early universe cosmology, condensed matter physics, and the black hole to nuclear physics, which has led to several advancements in pure mathematics. String theory can be seen as a potential "theory of everything." It aims to explain particle physics and gravity because it features a mathematical model that describes the fundamental forces and the forms of matter that exist in the universe.

The first study of string theory was done in the 1960s. The theory was then carried out to study the strong nuclear force. It was later abandoned with the introduction of quantum chromodynamics. As time went on, scientists discovered that the very characteristics that made this theory not fit for nuclear physics made it a potential answer for the quantum theory of gravity. The first version of string theory is called bosonic string theory, and it included just a type of particle called bosons. The theory later evolved into the superstring theory, which describes the link known as supersymmetry between the class of particles known as fermions and bosons. Subsequently, five separate consistent types of the superstring theory were proposed, and in the mid-90s, it was presumed that the five versions were all separate from the limited scenarios of one theory in 11 dimensions called M-Theory. Later, in 1997, an elementary relationship known as AdS/CFT correspondence was discovered by certain researchers. This relationship describes the link between string theory and another kind of theory known as quantum field theory.

String theory had many challenges, one of the major ones being that the definition of the complete string theory is satisfactory in all situations. It was also thought that the theory describes a host of potential universes, which made it difficult for researchers to develop particle physics theories with the string theory. These problems led several critics to argue if there is any need to continue pursuing the unification of string theory considering the confusing approaches.

Fundamentals of String Theory

Two theoretical frameworks were introduced in the previous century to form the laws of physics. The first attempt to do so was by Albert Einstein with his general theory of relativity. His theory aimed to describe the existence of space-time and gravity. The second theory is quantum mechanics, which aims to explain physical phenomena with the aid of known principles of probabilities. As of the late 1970s, both frameworks were enough to describe most of the observable events in the universe ranging from the evolution of stars and the whole universe to elementary particles and atoms.

Despite the success of both frameworks, many problems were still unexplainable. One such fundamental topic was that of quantum gravity. Einstein used classical physics in explaining his general theory of relativity. In contrast, the other framework of quantum mechanics described other fundamental forces. The world needed a quantum theory of gravity. The problem with doing so was that it was difficult to apply the quantum theory to explain gravity's force. Other issues dealt with black holes and atomic nuclei that are difficult to understand.

The goal of string theory was to address these issues as well as other problems. String theory began by assuming that we can model the point-like particles in particle physics as one-dimensional objects known as strings. These strings are propagated through space, and they interact with each other. One version of the theory describes just one type of string, which probably exists as a small loop of ordinary string and has different ways of vibration.

String theory has seen several different advances, with one of them being discovering some dualities. These dualities are mathematical transformations that can tell a physical theory from another. Scientists involved in studying string theories have found several of these dualities that exist between various versions of string theory. This led scientists to conclude that the different consistent versions of string theory are part of one framework called M-theory. Research into string theory has produced several results in areas like our interactions with gravity and the nature of black holes.

Scientists' attempts to understand the quantum areas of black holes have produced some paradoxes and our research into string theory aims to understand these problems. One of the big results of this research is the anti-de-Sitter/conformal field theory correspondence, also known as AdS/CFT, discovered in 1997. The AdS/CFt shows the relationship between other physical theories with string theory. It has also been applied to studying quantum gravity and black holes, nuclear physics, and condensed matter physics.

Scientists hope that string theory could answer all human questions about the universe, which will make it the theory of everything. This hope is because string theory brings together all the elementary interactions, which include gravity. The major challenge of string theory is that it fails to offer a satisfactory definition in all situations. We can define the scattering of strings by applying perturbation theory techniques which refers to a category of methods of analysis used for determining probable solutions of nonlinear equations when the precise solutions can't be gotten. They are used to predict, show, and describe the nature of vibrating systems occurring due to nonlinear effects.

Another unclear point in string theory is the principle that supports how it picks its vacuum state, which is the physical state responsible for determining the universe's properties.

The application of quantum mechanics to physical objects such as the electromagnetic field, space, and time, is known as quantum field theory. In particle physics, quantum field theories form the basis for our understanding of elementary particles. The perturbation theory was developed by Richard Feynman and other researchers in the 20th century. The theory applies unique diagrams known as Feynman diagrams for calculations. The perturbation theory is used to calculate the likelihood of several physical happenings in quantum field theory. According to research, these diagrams show the interactions of point-like particles and their paths.

Planck's length ($5.24934e^{-35}$ feet) is presumed to be the unique length scale of strings in several particle physics theories that depend on string theory. Planck's length is believed to be the point from where the impact of quantum gravity becomes significant. These types of objects are usually similar to zero-dimensional point particles on bigger scales like the visible scales in laboratories. In this case, one of the string's vibrational states will be consistent with the graviton.

CHAPTER 8:
QUANTUM PHYSICS IN YOUR DAILY LIFE

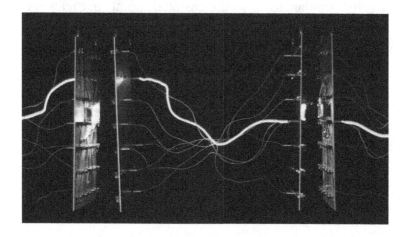

Whenever people hear the words "quantum physics," what comes to mind is something special and complicated that occurs in unique conditions in our universe. And maybe you have seen quantum computing in articles mentioning companies like IBM, Google, and Microsoft competing to be the first to develop a quantum computer.

Many people are not aware that quantum physics is part of our daily lives as much as eating. Once you are done reading this part of the book, you will realize how much quantum physics affects your daily life. Below are some of the ways quantum physics impact our daily lives.

Toaster

Your bread toaster is probably one of the most overlooked devices you use in your home. Imagine your life without your toaster; that crunchy bread you enjoy so much would be no more. While you sip your tea, have you ever thought about the science behind how your toaster works? Have you ever thought that quantum physics plays a huge role in the existence of your toaster?

Your toaster works by heating bread with the red glowing heating element of the toaster. It simply turns electrical energy into heat energy.

It's hard to miss the rows of bright red wires facing the bread when you look inside a toaster. In a similar way to hundreds of little radiators, these wires become heated when electricity runs through them.

Energy is transferred from one end of a wire to the other as electricity runs across it. Like water that flows down a conduit, so does the flow of energy. Electrons, the little atomic particles that constitute the wire, are responsible for transporting electrical energy along it. Flowing electrons crash with one other as well as other atoms, producing heat as they go through the metal wires. Additional collisions and heat are produced when the wire is thinner and the electricity is larger.

Many people consider toasters to be one of the major reasons for quantum physics. As the rod gets hotter, your toaster can toast your bread to the texture you want.

Fluorescent Light

Quantum physics is solely responsible for those brilliant lights you get from those beautiful curling bulbs. Fluorescent bulbs work by exciting a small amount of mercury vapor in the bulb. This results in the generation of the bright light you can see. So, you have quantum physics to thank for the way a fluorescent bulb works.

Computer and Mobile Phones

Imagine a world without a computer and your mobile phone. That sounds like the stone age. But this might be a possibility if quantum physics did not exist. The entire working mechanism of a computer is dependent on quantum physics. The computing world depends on electrons' movement, which behaves like a wave. This is the foundation of the creation of semiconductors which are extremely important in manufacturing computer and mobile phone devices. Our knowledge of the wave property of electrons is applied in manipulating the electrical characteristics of silicon.

Biological Compass

Humans are not the only living creatures to benefit from quantum physics. Some scientists have discovered that creatures such as the European robin depend on quantum physics for migrating from one-point t to another. Cryptochrome is a light-sensitive protein that contains electrons. As photons enter the bird's eye, they come into contact with the cryptochrome and cause the release of free radicals. This helps the bird to become aware of a magnetic map. Other theories claim that the bird's beak has magnetic minerals. Other animals that are thought to use a similar type of magnetic compass include lizards, crustaceans, insects, and certain mammals. We as humans also have a similar type of cryptochrome in our eyes, and while many think there may be a similar function in it, no one knows what it does for sure.

Transistor

The importance of transistors in modern life cannot be overestimated. They are generally used to change and amplify electric power and signals. When you take a closer look at what makes a transistor, you will realize that it is made up of several layers of silicon in addition to other components. Computer chips are made with millions of these types of transistors, and these chips are the engine house of most devices that have improved our lives massively. Without quantum physics, chips wouldn't exist, and neither would laptops, tablets, smartphones, and other gadgets that make your life so interesting.

Laser

Lasers work by applying the principles of quantum physics. Lasers work thanks to spontaneous emission, thermal emission, and finally, fluorescence. Electrons will go to a higher energy level when they are excited. But they won't remain at the higher level for long because they are in an unstable state, so to remain unstable, they will return to a lower energy level emitting light as a result. This is the basis of the workings of lasers.

Microscopy

Quantum physics has caused massive changes in how electron microscopes work. The quality of images of biological samples obtained using electron microscopes has improved massively. Therefore, researchers can get a larger amount of information from just one image.

Global Positioning System (GPS)

The invention of GPS is one of the most overlooked discoveries in this time and age. We always use it, that's why we give it for granted, and it has become such a huge part of our lives that without it we would—literally—feel lost. You can easily visit places you have never been before with the help of a GPS on your phone or in your car, and know how much time and distance it will take to reach your destination, which is calculated with the help of satellites working in unison to estimate the ETA given the real-time conditions of your trip. The satellites used for GPS have an in-built atomic clock that depends on quantum physics to work.

Magnetic Resonance Imaging

Another name for "magnetic resonance imaging" is "nuclear magnetic resonance." This device works by reversing the spin electrons present in hydrogen nuclei. This consists of a shift of energy which involves the application of quantum physics. MRI has made studying soft tissue extremely easier, and it has contributed to the quick diagnosis and treatment of certain fatal illnesses. Before the invention of the MRI machine, many of these diseases went undetected until they reached the fatal stages.

Telecommunication

The easy communication we have today is major because of our understanding of certain aspects of quantum physics. No one would have imagined two-way communications over thousands of miles as possible at some point in our history. But thanks to fiber-optic telecommunication, we can communicate with one another over thousands of miles. Fiber-optic communication works by using lasers, which were discovered thanks to quantum physics.

Optical fiber communication uses light via optical fibers to carry the signal from one point to another. As the optic fiber transmits the electric signal, it is converted to light at the receiving point, after which it is converted once again to an electrical signal. The data of the signal might be in audio, video, or any other transmittable form over long distances or on local networks.

CONCLUSION

This is the end of our journey. This guide had the goal to introduce you to some of the most interesting topics that surround the quantum physics arena. There is a whole universe behind each paragraph you have just read, and if you are interested in discovering it even more, I invite you to go deeper and specialize in your favorite topic.

But the general gist of how quantum physics works is no rocket science. Both classical physics and quantum physics aim to explain the science behind the universe. They seek answers to how and why the universe works the way it does.

It doesn't matter what you do for a living; understanding quantum physics can help you gain perspective on many things you've always given for granted in your life. But realizing that quantum physics is behind everything we are and is the very foundation of our being makes us no longer want to ignore it.

Quantum physics exists at the border between spirituality and science. It lends itself to many layers of reading, way far from what is normally comprehended in the realm of typical physical laws. It sparks reflections about other plans of existence and alternative realities, and maybe there will come a time and a place to talk about it.

It is our understanding of quantum physics that may be able to help us cross borders that today seem unthinkable and bring us to the very foundation of the universe. Questions concerning what our universe is and where it came from will finally be answered.

I hope this book has inspired you to search for the intrinsic beauty of our existence.

Made in United States
North Haven, CT
26 January 2024

47892546R00059